NAVY ENTERPRISES

EVALUATING THEIR ROLE IN PLANNING, PROGRAMMING, BUDGETING AND EXECUTION (PPBE)

JESSIE RIPOSO • IRV BLICKSTEIN
JOHN A. FRIEL • KINDLE FELL

Sponsored by the U.S. Navy

Approved for public release; distribution unlimited

NATIONAL DEFENSE RESEARCH INSTITUTE

The research described in this report was sponsored by the Assistant Deputy Director of the Navy's Programming Division (N8). The research was conducted in the RAND National Defense Research Institute, a federally funded research and development center sponsored by the Office of the Secretary of Defense, the Joint Staff, the Unified Combatant Commands, the Department of the Navy, the Marine Corps, the defense agencies, and the defense Intelligence Community under Contract W74V8H-06-C-0002.

Library of Congress Cataloging-in-Publication Data

Navy Enterprises: evaluating their role in planning, programming, budgeting and execution (PPBE) / Jessie Riposo ... [et al.].
p. cm.
Includes bibliographical references.
ISBN 978-0-8330-4664-2 (pbk. : alk. paper)
1. United States. Dept. of the Navy—Evaluation. 2. United States. Dept. of the Navy—Management—Evaluation. 3. United States. Dept. of the Navy—Appropriations and expenditures—Evaluation. I. Riposo, Jessie.

VB23.E83 2009
359.60973—dc22

2009003537

The RAND Corporation is a nonprofit research organization providing objective analysis and effective solutions that address the challenges facing the public and private sectors around the world. RAND's publications do not necessarily reflect the opinions of its research clients and sponsors.
RAND® is a registered trademark.

Cover design by Carol Earnest

© Copyright 2009 RAND Corporation

Permission is given to duplicate this document for personal use only, as long as it is unaltered and complete. Copies may not be duplicated for commercial purposes. Unauthorized posting of RAND documents to a non-RAND Web site is prohibited. RAND documents are protected under copyright law. For information on reprint and linking permissions, please visit the RAND permissions page (http://www.rand.org/publications/permissions.html).

Published 2009 by the RAND Corporation
1776 Main Street, P.O. Box 2138, Santa Monica, CA 90407-2138
1200 South Hayes Street, Arlington, VA 22202-5050
4570 Fifth Avenue, Suite 600, Pittsburgh, PA 15213-2665
RAND URL: http://www.rand.org/
To order RAND documents or to obtain additional information, contact
Distribution Services: Telephone: (310) 451-7002;
Fax: (310) 451-6915; Email: order@rand.org

Preface

The need to recapitalize the force structure while maintaining or improving fleet readiness in a fiscally constrained environment has led the Navy to adopt a new management paradigm: the Navy Enterprise construct. The purpose of the Navy Enterprise construct is to achieve efficiencies so that current and future readiness can be met with limited budgets. More specifically, the Navy Enterprise seeks to gain an improved return on investments through improved resource allocation and increases in output over cost. However, the exact implementation of Navy Enterprise is still being developed, and many questions remain unanswered. One such question is how should the Navy Enterprise be aligned with funding mechanisms? The Navy asked the RAND Corporation to assess the Navy Enterprise concept within the Planning, Programming, Budgeting and Execution (PPBE) framework.

This research was sponsored by the Assistant Deputy Director of the Navy's Programming Division (N8) and conducted within the Acquisition and Technology Policy Center of the RAND National Defense Research Institute, a federally funded research and development center sponsored by the Office of the Secretary of Defense (OSD), the Joint Staff, the Unified Combatant Commands, the Department of the Navy, the Marine Corps, the defense agencies, and the defense Intelligence Community. It should be of interest to persons concerned with the PPBE process, the Navy's organizational structure, or the Navy Enterprise concept.

For more information on this research, contact the principal authors of this report, Jessie Riposo, at riposo@rand.org, and Irv Blickstein, at irv@rand.org. For more information on RAND's Acquisition and Technology Policy Center, contact the Director, Philip Antón. He can be reached by e-mail at atpc-director@rand.org; by phone at 310-393-0411, extension 7798; or by mail at the RAND Corporation, 1776 Main Street, P. O. Box 2138, Santa Monica, California 90407-2138. More information about RAND is available at www.rand.org.

Contents

Preface ... iii
Figures ... vii
Tables ... ix
Summary ... xi
Acknowledgments ... xv
Abbreviations ... xvii

CHAPTER ONE
Introduction ... 1
Organization of This Report ... 4

CHAPTER TWO
The Navy Enterprise: Governance, Organization, and Other Elements ... 5
The Executive Committee ... 5
The Fleet Readiness Enterprise ... 7
The Providers ... 9
Governing Behaviors of the Navy Enterprise ... 11

CHAPTER THREE
A Description of the Planning, Programming, Budgeting and Execution Process and the Role of Enterprises ... 13
The PPBE Process and Associated Navy Organizations ... 13
Navy Enterprises Largely Participate in PPBE Through Resource Sponsors and Budget-Submitting Offices ... 16

Formal Participation of Warfare Enterprises and Providers Is Minimal, but Support Role Is Extensive 18
Interview Results Suggest That Participation of the Navy Enterprise in PPBE Has Had Limited Impact 23

CHAPTER FOUR
Alternative Constructs 27
No Involvement 27
Select Involvement 29
Process Ownership 31
Evaluation of Alternatives 33
Pros and Cons of Alternatives 36
No Involvement 36
Select Involvement 37
Enterprise Ownership 38

CHAPTER FIVE
Summary of Findings 41

Bibliography 43

Figures

1.1. Analytical Framework 3
2.1. Navy Enterprise Organizational Construct 6
2.2. Enterprise Maturity Levels 9
3.1. Planning, Programming, Budgeting and Execution Overlap 15
3.2. Enterprise Participation in the POM Process 24

Tables

3.1. Resource Sponsors and Responsibilities 17
3.2. Providers, Warfare Enterprises, Resource Sponsors, and BSO Relationships 19
3.3. Enterprise and Provider Participation in Navy's Primary PPBE Products 21
4.1. No Enterprise Participation 28
4.2. Current, No, and Select Enterprise Participation 30
4.3. Current, No, Select, and Ownership Participation 32
4.4. Evaluation of Alternatives Relative to Current Involvement 36

Summary

The Navy Enterprise has evolved over the past decade to achieve various objectives from improving efficiencies through lean, six-sigma efforts to producing the workforce of the future. As the objectives, goals, and structure of the organization have changed and grown, so has the very meaning of the Navy Enterprise. Currently, the Navy Enterprise is not only an organizational structure, but is a way of doing business, a behavioral model. However, the enterprise concept has been executed by a corresponding and evolving organizational structure. This organizational structure consists of a number of organizations, each having their own role, responsibilities, and functions in the Navy Enterprise.

This research is an evaluation of the participation of organizations within the Navy Enterprise in the PPBE system. The objectives of this research were to (1) identify and describe current participation of organizations in PPBE and (2) identify and evaluate potential alternatives for participation. RAND accomplished this through evaluations of available documentation and extensive interviews with nearly twenty senior leaders throughout the Navy.

Our investigations revealed that the formal role of Navy warfare enterprises and providers in PPBE has not changed much. The enterprises and providers mostly participate in the PPBE process via the various resource sponsors and budget submitting offices (BSOs),[1] as

[1] BSOs are the organizations that manage the databases containing the budget data. They submit the budgets to OSD for approval. BSOs include Manpower Personnel Training & Education, Naval Sea Systems Command, Naval Air Command, Space and Naval Warfare Systems Center, Naval Supply Systems Command, Naval Facilities Engineering Command,

they have in the past. The new activities that the enterprises and providers have participated in, such as the wedge liquidation and support provided to N1, were perceived to be beneficial.[2] The biggest benefit of the Navy Enterprise construct from a PPBE perspective has been the increased communication between resource sponsors, providers, and warfighters, which has helped the Navy to better assess cost and risk trade-offs for resource allocation decisions. However, the additional workload borne by the enterprises and additional complexity brought into the PPBE process could be greater than the benefit. The uncertain balance between costs and benefits resulted in interviewees being almost equally divided between those who thought the Warfare Enterprises' involvement should increase and those who thought it should decrease.

We identified and evaluated three alternative constructs for warfare enterprise and provider involvement in the PPBE process: (1) no involvement, (2) select involvement, and (3) PPBE process ownership. Together, these alternatives offer a full range of levels of participation for initial evaluation. We assessed the potential costs, benefits, and other considerations important for evaluating these alternatives using seven metrics. We assessed the total amount of workload required to execute an alternative, other potential costs of the alternative, potential PPBE benefits of the alternative, effect on alignment of the phases within PPBE, overall complexity, "buy-in" (or the sense of ownership for PPBE outcomes), and the ability to produce a Program Objective Memorandum (POM). In this system, the value of a metric is represented in terms of a four-point scale: Relative to current participation, costs or benefits of alternatives will be either better, slightly better, slightly worse, or worse.

Using this methodology, we did not identify any single preferred option. However, the current involvement serves as a good pilot for the

---

Commander Naval Installations Command, Office of Naval Research, Commander—Pacific Fleet, and Commander—Atlantic Fleet.

2 *Wedge liquidation* is the process of identifying cost reductions in order to achieve a balanced budget. This process is required when planned expenditures exceed the available budget.

development and evaluation of alternative constructs. Efforts should be made to foster the benefits of participation observed and to pursue ways to evaluate the cost of such participation. More broadly, many fundamental questions regarding the Navy Enterprise remain unanswered and should be the focus of future efforts. Specifically, answers to a number of questions: What is the purpose of the Navy Enterprise construct? Is the Navy Enterprise the correct approach to address the Navy's evolving goals? What organizations should be an enterprise? and What are the roles and responsibilities of enterprises?

Acknowledgments

The authors would like to thank all of the individuals who took time out of their busy schedules for our discussions and for the talent and knowledge of the RAND staff who helped to pull this work together. In particular, we would like to thank Nancy Harned, N80B, who helped us work through many of the tough issues we encountered over the course of this project. The content and conclusions presented in this work, however, remain solely the responsibility of the authors.

Abbreviations

ASN-RDA	Assistant Secretary of the Navy, Research, Development and Acquisition
BES	Budget Estimate Submission
BSO	Budget Submitting Offices
BUMED	Bureau of Medicine
CNIC	Commander Naval Installations Command
CNO	Chief of Naval Operations
DNS	Director, Navy Staff
EXCOM	Executive Committee
FMB	Office of Financial Management and Budget
FRAGORD	fragmentary order
FRE	Fleet Readiness Enterprise
IPVT	Investment Pricing Validation Team
JCS	Joint Chiefs of Staff
MPT&E	Manpower Personnel Training and Education
MPVT	Model Pricing Validation Team
N091	Director, Navy Test and Evaluation, Technical Requirements
N093	Surgeon General of the Navy
N095	Chief of Navy Reserve
N1	Deputy Chief of Naval Operations, Manpower and Personnel

N2	Director, Naval Intelligence
N3/N5	Deputy Chief of Naval Operations for Information, Plans and Strategy
N4	Deputy Chief of Naval Operations, Fleet Readiness and Logistics
N6	Deputy Chief of Naval Operations, Communication Networks
N8	Deputy Chief of Naval Operations for Integration of Capabilities and Resources
N8F	Deputy Chief of Naval Operations, Warfare Integration
N80	Director, Programming Division
N81	Director, Assessment Division
N82	Director, Fiscal Management Division
N84	Director, Oceanographer
N85	Director, Expeditionary Warfare Division
N86	Director, Surface Warfare Division
N87	Director, Submarine Warfare Division
N88	Director, Air Warfare Division
N89	Director, Special Programs Division
NAE	Naval Aviation Enterprise
NAVAIR	Naval Air Systems Command
NAVFAC	Naval Facilities Command
NAVSEA	Naval Sea Systems Command
NAVSUP	Naval Supply Systems Command
NCB	Director, Office of Budget and Reports
NE	Navy Enterprise
NECE	Navy Expeditionary Combat Enterprise
NETWARCOM	Navy Network Warfare Command
NNFE	Naval NETWAR/FORCEnet Enterprise

ONR	Office of Naval Research
OP-02	Deputy Chief of Naval Operations, Undersea Warfare
OP-03	Deputy Chief of Naval Operations, Surface Warfare
OP-05	Deputy Chief of Naval Operations, Naval Aviation
OPNAV	Office of the Chief of Naval Operations
OSD	Office of the Secretary of Defense
PEO	program executive office
PLANORD	planning order
POM	Program Objective Memorandum
PPBE	Planning, Programming, Budgeting, and Execution
QDR	Quadrennial Defense Review
SECNAV	Secretary of the Navy
SPAWAR	Space and Naval Warfare Systems Command
SSBN	Ballistic Missile Submarine, Nuclear
SSGN	Submersible Ship, Guided Missile, Nuclear
SSN	Attack Submarine, Nuclear
SWE	Surface Warfare Enterprise
TYCOM	type commander
UNSECNAV	Undersecretary of the Navy
USE	Undersea Warfare Enterprise
USFF	U.S. Fleet Forces
VCNO	Vice Chief of Naval Operations

CHAPTER ONE

Introduction

In industry, there are many definitions for what constitutes an enterprise. However, culling the common features of these definitions, an *industry enterprise* can be defined as an organization or group of organizations under the same ownership or control that deliver a product to a customer in return for a profit of some kind.

Though the enterprise concept is rooted in industry, the Navy's implementation of the concept differs from those in industry because the Navy has dissimilar goals and objectives from an industry enterprise. As Chief of Naval Operations ADM Gary Roughead put it: "I don't want to turn the Navy into a business, but we need to understand the business of the Navy." As the Navy has worked to adopt the enterprise concept, the purpose, meaning, and implementation of the concept for the Navy has evolved. It continues to evolve today. In March 2008, the Navy Enterprise construct was described by Navy Enterprise Chief Operating Officer RADM David Buss as

> activities, governance, and behaviors that will drive additional efficiencies in how the Navy delivers current readiness and future capability, as well as provide a foundation for making better, more informed mission, capability, resource allocation, and risk decisions.[1]

Today, the purpose of the Navy Enterprise construct, as stated by the same source, is to achieve additional efficiencies so that current and future readiness can be met with limited budgets. More specifically,

[1] Navy Enterprise, "Navy Enterprise," Web site, undated.

the Navy Enterprise seeks to gain an improved return on investments by "improving output over cost" and "improving resource allocation effectiveness."[2] These are broad objectives, for which many initiatives and specific action plans outside of the enterprise itself will be required. Though the enterprise framework might help to organize the initiatives, responsibilities, and authority for actions required to achieve these objectives, it is not intended to serve as the solitary vehicle for achieving these goals.

Because of the evolving nature of this maturing concept and the lack of documentation and guidance, there is still a great deal of confusion regarding the "who, what, and how" of the Navy Enterprise. A recent study by the Executive Leadership Group, Inc., identified many sources of confusion, including uncertainty as to what and who a enterprise is, what the goals and metrics of an enterprise are, what the roles and relationships are, and how funding mechanisms align to the new construct.[3] To help clarify some of this confusion, RAND was asked by the Director, Programming Division (N80) to

- review the Navy Enterprise construct within a Planning, Programming, Budgeting and Execution (PPBE) context
- explore alternative constructs to support the Navy's PPBE goals and objectives.

RAND conducted a series of interviews and document evaluations to identify the goals and objectives of PPBE and the Navy Enterprises, identify roles and responsibilities, define current Navy Enterprise participation in PPBE, and collect information on the availability of data and metrics that can be used to evaluate alternative constructs. Figure 1.1 depicts the analytical framework RAND employed to answer the research questions.

---

[2] Navy Enterprise, "Introduction to Navy Enterprise," briefing, undated, slide 1. In previous incarnations of the Navy Enterprise concept, as recent as April 2007, the objectives have included generating sufficient buying power, creating the 21st century workforce of the Navy, and maintaining or improving current levels of readiness and military effectiveness.

[3] Wendi Peck, *Reaching Across Boundaries to Achieve the Right Results: A Framework for Enterprise Workshop Development*, The Executive Leadership Group, Inc., June 2007.

Figure 1.1
Analytical Framework

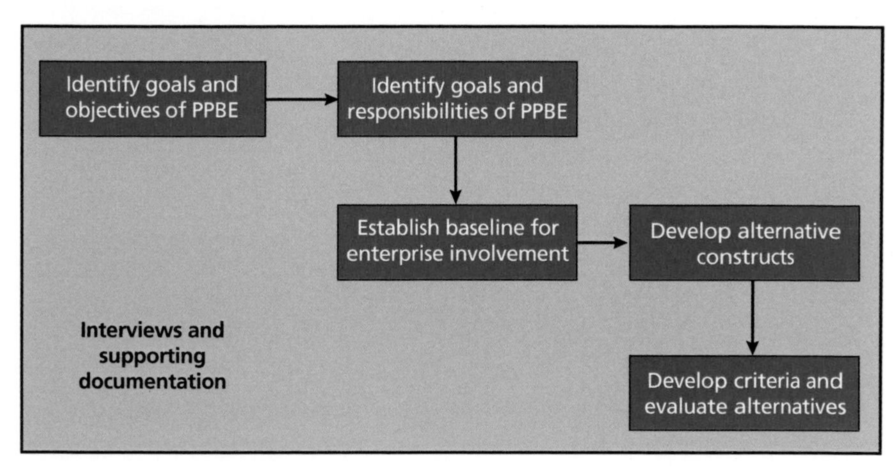

RAND *MG794-1.1*

Interviewees included a variety of representatives from the Navy Enterprise and the PPBE community:

- Deputy Director, Assessments Division (N81B/N00XB)
- Deputy Director, Warfare Integration (N8FB)
- Head, Program Planning and Development Branch (N801)
- Head, Sea Pillar Branch (N801J)
- Deputy Director, Programming Division (N80)
- Deputy Commander, Fleet Forces Command
- Director, Warfare Integration
- Director, Surface Warfare Division (N86)
- Deputy Director, Submarine Warfare (N87)
- Deputy Director, Expeditionary Warfare Division (N85)
- Director, Air Warfare (N88)
- Executive Director, Naval Air Systems Command (NAVAIR)
- Commander, NAVAIR
- Commander, Naval Sea Systems Command (NAVSEA)
- Assistant Deputy Director, Fleet Readiness and Logistics (N4B)
- Deputy Chief of Naval Operations for Manpower, Personnel, Education and Training/Chief of Naval Personnel (N1)

- Assistant Deputy for Manpower, Personnel, Education & Training/Chief of Naval Personnel (N1B).

RAND analysts established a baseline of enterprise participation from the current Program Objective Memorandum, Fiscal Year 2010 (POM-10) Planning Order (PLANORD)[4] and developed alternatives to this construct based on information obtained from interviews and from RAND Corporation subject matter experts. The pros and cons of alternative constructs are identified for evaluation by decisionmakers, although the evaluation criteria were constrained by available data.

Organization of This Report

Chapter Two provides an overview of the Navy Enterprise. Chapter Three outlines the goals and responsibilities of the PPBE process, defines how the Navy Enterprises currently participate and introduces the alternative approaches that we identified. Readers familiar with PPBE and the Navy Enterprise construct can comfortably proceed directly to the end of Chapter Three (p. 23), which begins the description of our findings. Chapter Four summarizes the evaluations of the alternatives and highlights additional considerations. Chapter Five offers a brief summary and recommended next steps.

[4] The PLANORD is a formal memorandum from the Vice Chief of Naval Operations that provides instructions and guidance for programming an executable Navy budget.

CHAPTER TWO

The Navy Enterprise: Governance, Organization, and Other Elements

As illustrated in Figure 2.1, the Navy Enterprise consists of three main entities: an Executive Committee (EXCOM); the Fleet Readiness Enterprise; and a group of providers of resources and services. In this chapter, we describe the role of each entity and identify several issues relevant to participation of the Navy Enterprise in PPBE, such as the maturity of different Navy Enterprise entities and which ones control resources. We also discuss the information systems and tools that support the Navy Enterprise, as well as organizational behaviors and cultural changes the Navy considers key components of the enterprise construct.

The Executive Committee

The EXCOM is the governing body responsible for setting Navy Enterprise objectives, evaluating Navy Enterprise output and progress, supporting Navy Enterprise progress (by removing barriers), making decisions (resource allocation, budgeting, and others), and developing strategic communications. The EXCOM is comprised of the senior leadership of the Navy: the Secretary of the Navy; the Chief of Naval Operations; the Assistant Secretary of the Navy for Research, Development, and Acquisition; the Vice Chief of Naval Operations; the Commander, Fleet Forces Command; the Director, Navy Staff; the Navy Enterprise Chief of Staff; the Director, Programming Divi-

Figure 2.1
Navy Enterprise Organizational Construct

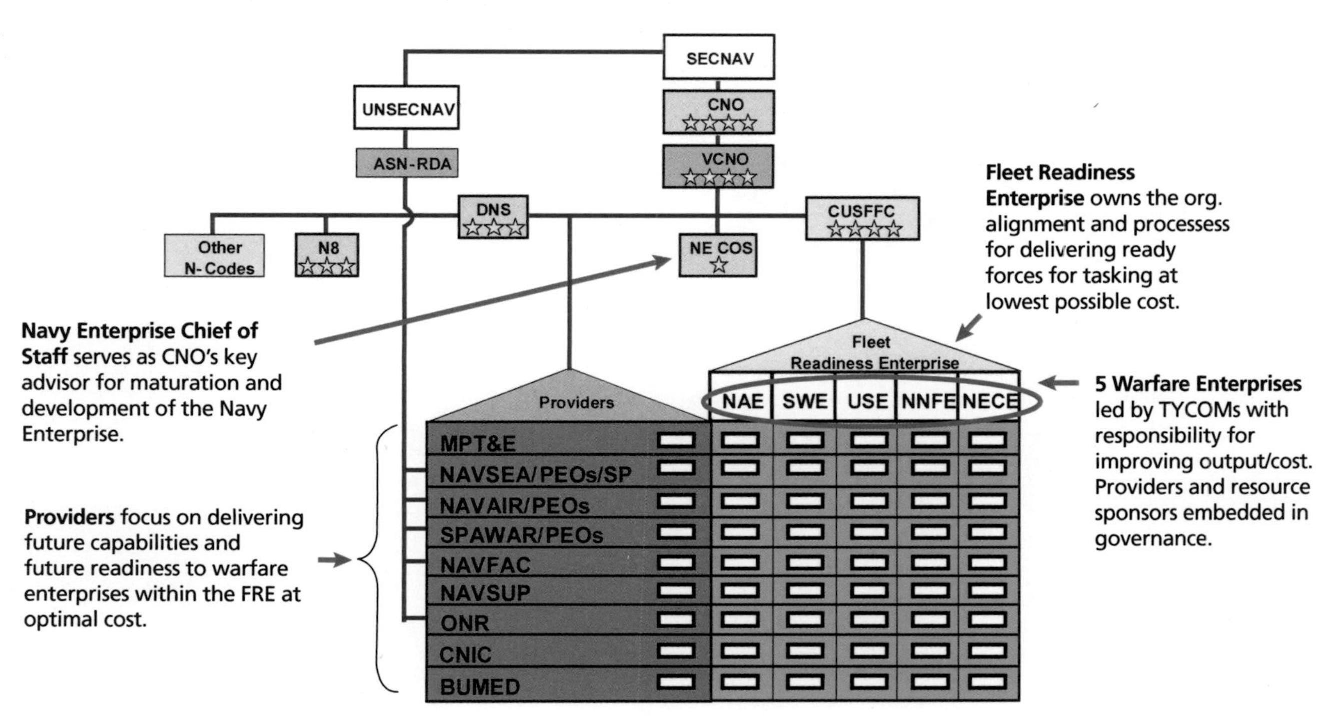

RAND *MG794-2.1*

sion (OPNAV N8); and the Assistant Secretary of the Navy, Financial Management and Comptroller.

The Fleet Readiness Enterprise

The Fleet Readiness Enterprise is responsible for delivering total force readiness at the lowest possible cost. It is led by the Commander, United States Fleet Forces Command, and consists of five warfare enterprises:

- Naval Expeditionary Combat Enterprise
- Surface Warfare Enterprise
- Undersea Warfare Enterprise
- Naval NETWAR/FORCEnet Enterprise
- Naval Aviation Enterprise.

Each warfare enterprise is responsible for identifying ways to improve output over cost.

All Navy warfare enterprises have the same goal: to increase efficiency and effectiveness. But they achieve these goals in different ways, and they vary in their maturity levels, as can be seen in the following brief descriptions of the leadership and mission of each enterprise:

- The Commander, Naval Submarine Force (Atlantic) is the head of the Undersea Warfare Enterprise, which consists of all stakeholders and resources supporting or operating SSN, SSGN, SSBN, fixed surveillance, or mobile surveillance forces.[1] The Undersea Warfare Enterprise focuses on increasing effectiveness and efficiency by improving the operational availability of the submarine fleet, improving commanding officer decisionmaking, ensuring the presence of experienced submarine personnel throughout the defense community, and generating the capability required to maintain undersea superiority in the future.

[1] Commander, Submarine Force, "Undersea Enterprise (USE) Overview," undated.

- The Naval Aviation Enterprise is led by the Commander, Naval Air Forces. The mission of the Naval Aviation Enterprise is to support the warfighter by providing combat-ready Naval Aviation forces. The Naval Aviation Enterprise measures efficiency and effectiveness by a single metric: aviation units ready for tasking at reduced cost, which is accomplished by improved reliability, process efficiencies, reduced cycle time, and other efforts.
- Like the Naval Aviation Enterprise, the Surface Warfare Enterprise mission is to provide combat-ready surface warfare forces to the fleets and combatant commanders. The Surface Warfare Enterprise is led by the Commander, Naval Surface Forces. The Surface Warfare Enterprise also measures its efficiency and effectiveness by warships ready for tasking.
- The NETWAR/FORCEnet Enterprise consists of commands involved in the business of command, control, communications, computers, collaboration, and intelligence and information operations, such as Space and Naval Warfare Systems Command (SPAWAR) and the Program Executive Office (PEO) for Command, Control, Communications, Computers, and Intelligence. The mission of the NETWAR/FORCEnet Enterprise is to provide and operate a global network to win battles in the Information Age. It is led by the Commander, Navy Network Warfare Command.
- Finally, the Navy Expeditionary Combat Enterprise—which consists of the enterprise's staff, all subordinate commands, and all other commands that influence and support the warfighting capability of the enterprise[2]—is also currently establishing processes and behavioral constructs to achieve greater efficiency and reduce costs and plans to develop metrics subsequently. This enterprise is led by the Commander, Navy Expeditionary Combat Command. It provides a number of services, including Explosive Ordinance Disposal, Diving Operations, Naval Construction, and expeditionary training.

---

2 Navy Expeditionary Combat Command, "About the NECE," Web page, undated.

The enterprises have varying maturity levels (see Figure 2.2). Some grew out of previously existing organizations, such as the Naval Aviation Enterprise, which has been in existence in some form for nearly nine years. Others are newly established, such as the NETWAR/FORCEnet Enterprise, which has been in existence for nearly two years. Consequently, some enterprises have a more established infrastructure and operational procedures that more easily facilitate participation in PPBE.

The Providers

In the Navy Enterprise construct, providers are responsible for providing services, equipment, and other resources to the Warfare Enterprises (and to the Navy overall). The providers' focus is on ensuring

Figure 2.2
Enterprise Maturity Levels

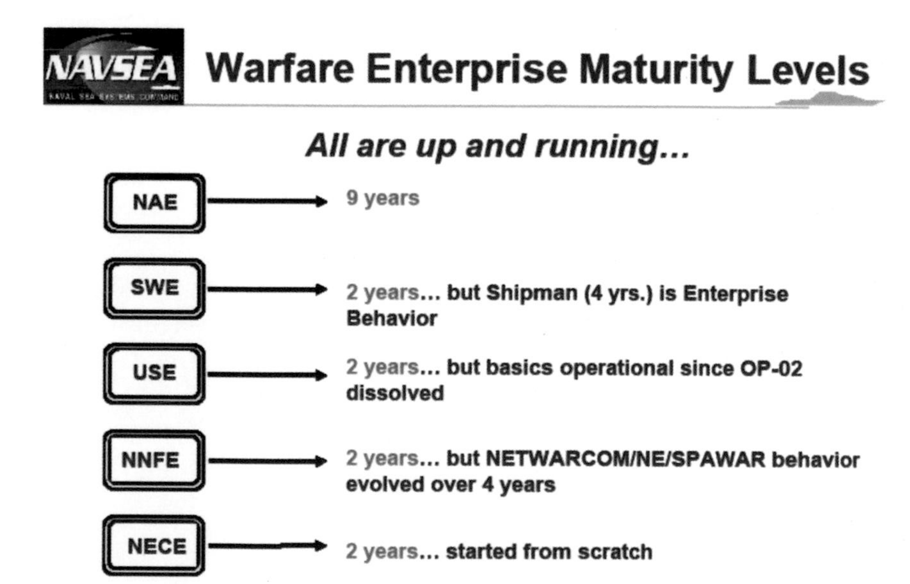

SOURCE: VADM Paul E. Sullivan, Commander, Naval Sea Systems Command, "Navy Enterprise Background, Evolution and Life in the Navy Enterprise," briefing, New Flag Officer Traning Symposium, August 2007.
RAND *MG794-2.2*

future readiness across the warfare enterprises at a minimal cost. There is overlap between the providers and warfare enterprises. For example, some individuals that are part of the NAVAIR (a provider) are also part of the Naval Aviation Enterprise (a warfare enterprise). The impetus for incorporating providers into the construct was the desire to achieve the same efficiencies and cost savings expected from the warfare enterprises. Other events also served as an impetus, such as a 2005 tasker from the Chief of Naval Operations directing that NAVAIR, NAVSEA, Naval Supply Systems Command (NAVSUP), SPAWAR, and Naval Facilities Command (NAVFAC) fall under a single systems command. The Navy decided that all five organizations would develop a common behavioral model rather than execute a structural change.[3] The behavioral model and enterprise construct was then extended to various other organizations in an effort to operate and manage all components of the budget under the same management paradigm.

Currently, there are nine providers:

- Bureau of Medicine (BUMED)
- Office of Naval Research
- NAVFAC
- SPAWAR and associated program executive offices
- NAVSUP
- NAVSEA and associated PEOs
- Commander, Naval Installations Command
- Manpower Personnel Training and Education (MPT&E)
- NAVAIR and associated PEOs.

These provider organizations control parts of the funding required to support the needs of the warfare enterprises. For example, MPT&E has budget authority and management responsibility for military manpower required by the warfare enterprises and other providers. The warfare enterprises communicate their manpower and training needs

[3] The behavioral model refers to efforts and actions to improve efficiency and reduce costs. In addition, personnel are encouraged to avoid myopic thinking and focus on what is best for the Navy as a whole and not only what is best for their particular command or organization.

to MPT&E, and it evaluates these requirements across the Navy to determine manpower levels.

Governing Behaviors of the Navy Enterprise

No description of the Navy Enterprise would be complete without mention of the *behaviors* touted as paramount to the success of the Navy Enterprise construct. Primary behaviors include the organizational discipline to control demand by requesting only that which is needed and no more, accountability, and developing the trust required in the sharing of data and information across organizations for better decisionmaking.

CHAPTER THREE

A Description of the Planning, Programming, Budgeting and Execution Process and the Role of Enterprises

The Navy Enterprise's participation in the Department of Defense's resource allocation process—the PPBE system—can be described by phase of participation (when), the activities or products participated in (what), and the level of or role in participation (how). To address these aspects, we begin by identifying the objectives of each phase in the PPBE process and the key Navy organizations associated with each phase. We also examine the relationship between Navy Enterprises and Office of the Chief of Naval Operations (OPNAV) resource sponsors and Navy budget submitting offices (BSOs). OPNAV and BSOs play key roles in the planning, programming, and budgeting phases of PPBE. Using the POM-10 PLANORD, we then identify the PPBE phases in which the Navy Enterprise most heavily participates, the extent and nature of its participation, and its specific contributions and inputs to the process. We then use current participation as a baseline against which we assess alternative constructs for Navy Enterprise involvement in PPBE.

The PPBE Process and Associated Navy Organizations

A group of organizations in OPNAV are responsible for ensuring that the programming objectives of PPBE are met. The core objectives, products, and entities associated with of each phase are as follows:

- **Planning.** In the planning phase, the objective is to identify the capabilities needed to satisfy national security requirements now and in the future. The Office of the Secretary of Defense (OSD) and the Joint Chiefs of Staff (JCS) publish strategic and planning guidance,[1] which the Deputy Chief of Naval Operations for Information, Plans and Strategy (N3/N5) and the Navy Director, Assessments and QDR (N81), use in generating Navy-specific planning guidance for the next phase in the PPBE process.
- **Programming.** In this phase, the objective is to array resources against defense programs intended to meet validated capability needs; the optimal balance among programs is sought to effectively manage risk and cost variables. The programming phase is managed by the individual components—in OPNAV, the Director, Programming (N80)—and overseen by N8 (Deputy Chief of Naval Operations for Integration of Capabilities and Resources). The N8 builds the program for the Chief Naval Officer to present to the Secretary of the Navy. Multiple resource sponsors also have responsibilities for programming in this phase.[2] The main products of the programming phase are the POM and subsequent Budget Estimate Submission (BES), which are reviewed by OSD and JCS.
- **Budgeting.** The objective of this phase is to develop the President's budget by arraying resources into an appropriate appropriations account structure and present an appropriation request to Congress. In the budgeting phase, the Office of Financial Management and Budget (FMB) is responsible for balancing the Department of the Navy budget across the Navy and the Marine Corps and present an executable program to the Secretary of the Navy.

[1] The Strategic Planning Guidance, Defense Fiscal Guidance, the National Military and National Security Strategies, the Quadrennial Defense Review (QDR), the Joint Planning Document, the Joint Programming Document, the Integrated Priority List (combatant commanders high-priority needs), and others.

[2] OPNAV organizations responsible for different programs include N1 (Manpower and Personnel), N4 (Fleet Readiness and Logistics), N6 (Communication Networks), N091 (RDT&E, S&T), N84 (Oceanography), N85 (Expeditionary), N86, N87, N88, and N89 (Special Programs).

To formulate and justify the Navy budget, FMB, the Director, Fiscal Management Division (N82), and the BSOs use the POM and BES documents developed during the programming phase.

- **Execution.** In this phase, the objective is to effectively manage appropriated funds to accomplish budgeted workload. Once the budget has been passed by Congress and an appropriations bill has been signed, it is the responsibility of the BSOs and fleets to execute the budget.

In a single year, all four phases are occurring simultaneously, as shown in Figure 3.1. In 2008, for example, the execution of the 2008 budget was underway. Congress was also reviewing the budget for the following year (2009). The planning and programming activities for 2010 and 2011 were beginning.[3] Toward the end of 2008, further planning for 2011 was occurring. In 2009, the same activities will be under-

Figure 3.1
Planning, Programming, Budgeting and Execution Overlap

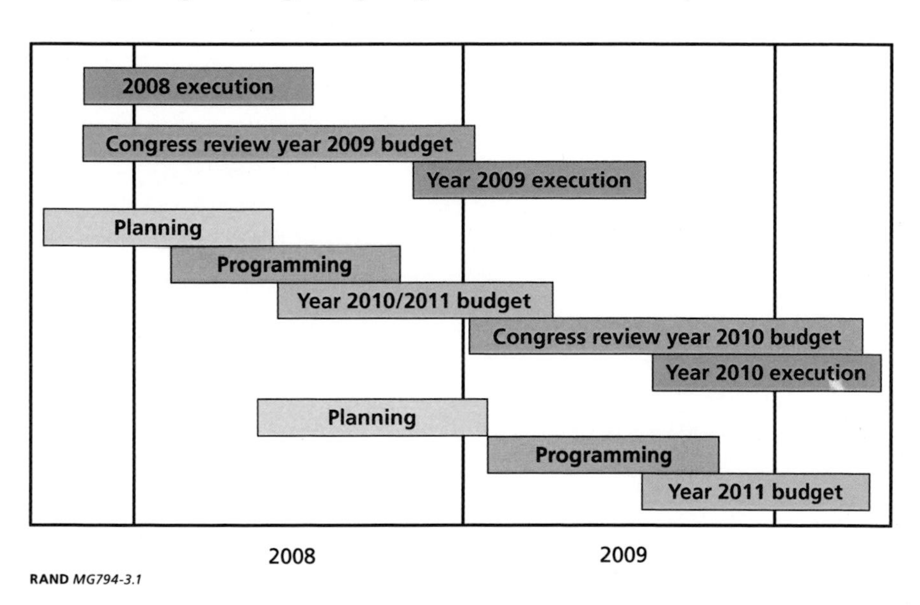

RAND *MG794-3.1*

[3] In even-numbered years (such as 2008) the budget is put together for two years (2010 and 2011). In odd-numbered years, such as 2009, the budget is put together for only a single year (2011).

taken. However, in odd-numbered years, the budget is put together for a single year only. In 2009, the budget for 2011 will be revised and updated. The cycle will repeat similarly for odd- and even-numbered years.

Navy Enterprises Largely Participate in PPBE Through Resource Sponsors and Budget-Submitting Offices

Analyses of the warfare enterprises' and providers' participation in PPBE is complicated by relationships with two organizations that have a central role in the process: resource sponsors and BSOs. In the programming phase, resource sponsors are the primary organizations responsible for allocating funds required to execute various programs. The warfare enterprises often voice their preferences and concerns to the resource sponsors. In some cases the N88, N87, N86, or N85 resource sponsor represents their respective enterprise in its submissions to the PPBE process. In other cases, such as with N1, the enterprises provide solicited feedback to guide their programming decisions. This implies that the enterprises participate in PPBE through their resource sponsors. BSOs manage the fiscal databases and submit the budgets to FMB, which then submits the budget to OSD for approval. Many of the providers are also BSOs, which can further complicate analysis of Navy Enterprise participation in PPBE.

Each resource sponsor is a member of the OPNAV staff and is responsible for a specific group of resources. For example, the Personnel Resource Sponsor (N1) will array resources required to supply military staff for all programs in the Navy, including those required by the system commands and PEOs.[4] Table 3.1 lists all of the resource sponsors and the resources for which they are responsible.[5]

---

[4] There are five system commands responsible for providing products and services to the fleets: Naval Sea Systems Command, Naval Supply Systems Command, Space and Naval Warfare Systems Command, Naval Air Systems Command, and Navy Installations Command.

[5] The Marine Corps has a structure that is separate and unique from the OPNAV staff.

Table 3.1
Resource Sponsors and Responsibilities

Resource Sponsor	Resource Responsibility
Deputy Chief of Naval Operations for Manpower and Personnel (N1)	Navy Total Force (MPT&E)
Director, Naval Intelligence (N2)	Intelligence functions
Deputy Chief of Naval Operations, Plans, Policy and Operation (N3/N5)	Planning and policy development
Deputy Chief of Naval Operations, Fleet Readiness and Logistics (N4)	Fleet and ashore readiness and logistics
Deputy Chief of Naval Operations, Communication Networks (N6)	Space command and control network communications and information functions
Deputy Chief of Naval Operations for Integration of Capabilities and Resources (N8)	Establish warfighting requirements and integrate all Navy resources
Director, Oceanographer (N84)	Oceanography, navigation
Director, Expeditionary Warfare (N85)	Expeditionary forces
Director, Surface Warfare (N86)	Surface forces
Director, Submarine Warfare (N87)	Submarine forces
Director, Air Warfare (N88)	Aircraft carriers, Air forces
Head, Special Programs (N89)	Special programs
Director, Navy Test and Evaluation, Technical Requirements (N091)	Navy science and technology, test and evaluation for all Navy programs
Surgeon General of the Navy (N093)	Medical
Chief of Navy Reserve (N095)	Reserve forces requirements

The warfare enterprises participate in PPBE by submitting their requirements and priorities to various resource sponsors; also, the enterprise leads can and do use their resource sponsors as a conduit for resolving POM issues. Additionally, many of the resource sponsors engage the warfare enterprises and the providers when putting together their budgets. For example, in putting budgets together, N1 will receive a baseline estimate of the personnel requirements from each warfare enterprise, including estimates of the dollars being consumed and the capabilities being delivered by personnel. In addition, N1 might ask the enterprises to evaluate where to make cuts if a reduction in manpower is needed. For example, if a reduction in the MPT&E budget is required, the enterprises may determine that more civilian labor or contractors should be used in lieu of uniformed staff. This decision construct can, however, have unintended consequences. For example, one resource sponsor RAND interviewed indicated that the manpower cuts made by one enterprise had adversely affected the cost and schedule of a particular procurement program.

Providers have the same number of resource sponsors as the enterprises, and both have primary as well as secondary sponsors, as shown in Table 3.2. Providers participate in PPBE through resource sponsors by identifying the costs associated with each resource sponsors' program. To explain further, the services and material that are required for various resource sponsors' programs, and offered by providers, are paid for by the resource sponsors. Another difference between the warfare enterprises' and providers' participation in PPBE is that most providers also participate in PPBE through their additional functions as BSOs (Table 3.2 identifies providers that are also BSOs). This means that the provider organization submits its own budget. The budgets of the warfare enterprises are submitted by the Fleet Readiness Enterprise leads and associated resource sponsors.

Formal Participation of Warfare Enterprises and Providers Is Minimal, but Support Role Is Extensive

The components of the Navy Enterprise have historically participated in the execution and budgeting (through the BSOs) phases of PPBE. The emphasis of the Navy Enterprise's participation in PPBE has been on execution. More recently, the enterprises have taken on a new role in the PPBE process by participating in the programming phase of PPBE. They participate in the programming phase by supporting other organizations (mostly resource sponsors) to accomplish certain activities or they generate inputs to the process themselves (the *what*). The role of

Table 3.2
Providers, Warfare Enterprises, Resource Sponsors, and BSO Relationships

	Organization	BSO	Primary Resource Sponsor	Secondary Resource Sponsors
Providers	MPT&E	Yes	N1	N091, N1, N4, N61, N85, N86, N87, N88, DNS
	NAVSEA	Yes	N86	N091, N2, N61, N84, N85, N86, N87, N88, N8F, DNS
	NAVAIR	Yes	N88	N091, N1, N2, N4, N61, N81, N82, N85, N86, N87, N88, N89
	SPAWAR	Yes	N6	N1, N2, N4, N81, N82, N84, N85, N86, N87, N88, DNS
	NAVSUP	Yes	N4	N61, N85
	NAVFAC	Yes	N4	N61, N85
	BUMED	Yes	N093	N4, N61, N85
	CNIC (Installations)	Yes	N4	N61, N82, N86, N87, N88
	ONR	Yes	N091	N1, N2, N4, N61, N86, N87, N8F
Individual warfare enterprises	SWE		N68/N85	N1, N4, N8F
	USE		N87	N1, N4
	NAE		N88	N1, N4
	NFFE		N6	N1, N4
	NECE		N85	N1, N4, N86, N8F
Fleet Readiness Enterprise	CFFC	Yes	N4	N1, N2, N61, N82, N84, N85, N86, N87, N88, N8F, DNS
	PACFLT	Yes	N4	N2, N61, N85, N86, N87, N88, N8F, DNS

enterprise and provider participation in PPBE activities is primarily a support role (the *how*). The greatest level of enterprise and provider participation in the POM-10 PPBE process occurred in developing options for wedge liquidation,[6] a process in which future expenses are reduced to maintain a balanced budget.[7] The Initial Wedge Liquidation Proposal was a consolidation of various enterprises' and providers' recommendations for reducing expenditures. NAVSEA/N4 then consolidated or pulled together the recommendations, and N8 reviewed the recommendations.

Table 3.3 lists a series of assessments and documents that were produced by the Navy in support of the PPBE process. These assessments and documents are described in detail in the next paragraph. The table shows that enterprise and provider participation in PPBE has increased over time. Enterprises and providers were not directed to participate until the PR 2009 cycle, and their participation increased in the POM 10 cycle. The level of support that the enterprises and providers lent to each assessment or document varies. In some cases, the production of these documents was heavily supported by the warfare enterprises and providers. As mentioned above, the greatest level of participation occurred in the wedge liquidation. In other cases, the warfare enterprises and providers provided very little support. For example, the enterprises and providers did not participate in the POM-10 PLANORD. This is the last document listed under the planning phase in Table 3.3 and is a formal memorandum from the Vice Chief of Naval Operations that provides instructions and guidance for programming an executable Navy budget.[8] It identifies who will participate in POM development, their associated responsibilities, a timeline, and deliverables required for creation of a POM. While the enterprises did not

---

[6] The wedge liquidation activity occurred in POM-10, but not in POM-08 and POM-09.

[7] During Program Review (PR) 09 the planned Navy's expenditures exceeded the Navy's authorized Top Line Obligational Authority, creating a "wedge." PRs are similar to POMs, but occur in odd-numbered years and cover only a 5-year time period.

[8] In the POM-10 planning phase, the Director of Naval Intelligence (N2) coordinates with commanders to provide an update on the threat environment. The Deputy CNO for Information, Plans and Strategy (N3/N5), submits Strategic Guidance. The CNO (through N3/N5) submits Strategic Guidance, and the Director, Assessments (N81) performs a gap analysis that projects areas of concern that could be ameliorated in the POM cycle. At the end of the POM cycle, the N81 determines how many gaps were resolved in the POM process. The last product delivered in the POM-10 planning phase is the POM-10 PLANORD.

Table 3.3
Enterprise and Provider Participation in Navy's Primary PPBE Products

Process	Product (Current Lead)	POM 08	POM 09	POM 10
Planning	Intelligence Update (N2)			
	Navy Strategic Guidance (N3/N5)		X	X
	Initial Wedge Liquidation Proposal (N4)*	NA	NA	X
	Front-End Assessment (N81)		X	X
	POM-PLANORD (N80)		X	X
Programming	Recommended Liquidation Actions (N4)*	NA	NA	X**
	Integrated Sponsor Capability Plan (N1/N6)			X
	Integrated Capability Plan (N81)			X
	Readiness Assessment (N4)			X
	Warfighting Capability Plan (N81)			X
	Initial Fiscal Guidance (FRAGORD) (N80)		X	X
	Sponsor Program Proposals (RS)			X
	Integrated Program Assessment (N81)			X
	POM Program Submission (N80)		X	X
Budgeting	BES (NCB)		X	X
Execution	Ready Force (USFF/BSO)		X	X

* Unique PON 10 activity.
** Provider only.

participate in the development of the document, it indicates that the enterprises and providers give "support" to all activities during the programming phase, with the exception of the N2 Intelligence Update.

Information about the amount of input received from or implemented by the Navy Enterprise in the other planning activities listed in Table 3.3 is limited and requires further investigation.

As indicated in Table 3.3, we identified nine major activities in the programming phase, and the Navy Enterprise participated in all of them during the POM 10 cycle. As in the planning phase, the warfare enterprises and providers participated to varying degrees in each activity.

The recommended liquidation action had the greatest amount of enterprise and provider participation and is the result of the planning phase liquidation drill discussed previously. In this drill, the enterprises and providers were instructed to find areas where expenditures or costs could be reduced. A representative of one provider interviewed by RAND mentioned they had found $2.4 billion in savings for the wedge liquidation.

The Integrated Sponsor Capability Plans, created by N1 and N6, identify resources required to provide a specific level of capability. Per direction from the Vice Chief of Naval Operations, N1 and N6 were instructed to collaborate with the warfare enterprises and providers: N6 was to provide a POM-10 assessment (an evaluation of the program) and N1 was to ensure the Navy's manpower aligns with requirements.

The Warfighting Capability Plan, created by N8F, identifies the warfighting capability required to balance risk (operational and financial) within the guidelines of the Strategic Planning Guidance.

The N81 Integrated Capability Plan (ICP) serves as an outline for Navy programming. It identifies programs and other investment priorities for achieving desired operational capabilities. The POM-10 PLANORD states that warfare enterprises and providers are to provide support in developing the ICP.

After capability assessments are complete, Initial Fiscal Guidance is issued by N80 to guide organizations in their development of programs and budgets.

The sponsor program proposal is used to identify and program strategy outlined in the ICP and fiscal guidance. Since many resource sponsors generate sponsor program proposals, a consolidated program proposal that integrates all areas is approved by N8 and presented to the Chief of Naval Operations as a tentative POM for review. The Chief of Naval Operations will review, change, and eventually approve this document. The development of resource sponsors' program proposals is also supported by warfare enterprises and providers.

At the end of the programming phase, N81 conducts an integrated program assessment, which is an assessment of the entire Navy program and its ability to meet requirements within the expected budget. Again, the warfare enterprises and providers support this assessment. The purpose of the assessment is to have an integrated evaluation of risk and requirements across resource sponsors and programs.

Once the integrated program assessment is complete and is reviewed and approved by the Chief of Naval Operations and the Secretary of the Navy, the POM-10 is submitted to OSD by N80.

The warfare enterprises and providers participate in fewer activities in the budget and budget execution phase—they support the development of the budget and the POM and ultimately provide the services and materials budgeted for, and required to, produce a ready force.

Figure 3.2 illustrates the extent of the warfare enterprises' and providers' participation in the major milestones in the POM-10 process and also provides a timeline of the above described events. The figure shows that only two activities (in the bright yellow boxes) are a primary responsibility of the warfare enterprises and providers in the process; however, the enterprises are providing support for the development of many of the deliverables in the POM process (the light yellow boxes).

Interview Results Suggest That Participation of the Navy Enterprise in PPBE Has Had Limited Impact

Interview results with seventeen senior officials in the Navy Enterprise and PPBE process indicate that the warfare enterprises and providers participate in the PPBE process mostly in a support role, as warfighters have in the past—that is, decision authority has remained in the hands of those organizations that have programming and budgeting responsibilities, the resource sponsors and BSOs. Interviewees were almost equally divided between those who thought the warfare enterprises' involvement should increase and those who thought it should decrease. However, the majority of interviewees favored eliminating the provider enterprise as a distinct entity within the Navy Enterprise. The reasons cited were a lack of commonality between providers, no clear decision-

Figure 3.2
Enterprise Participation in the POM Process

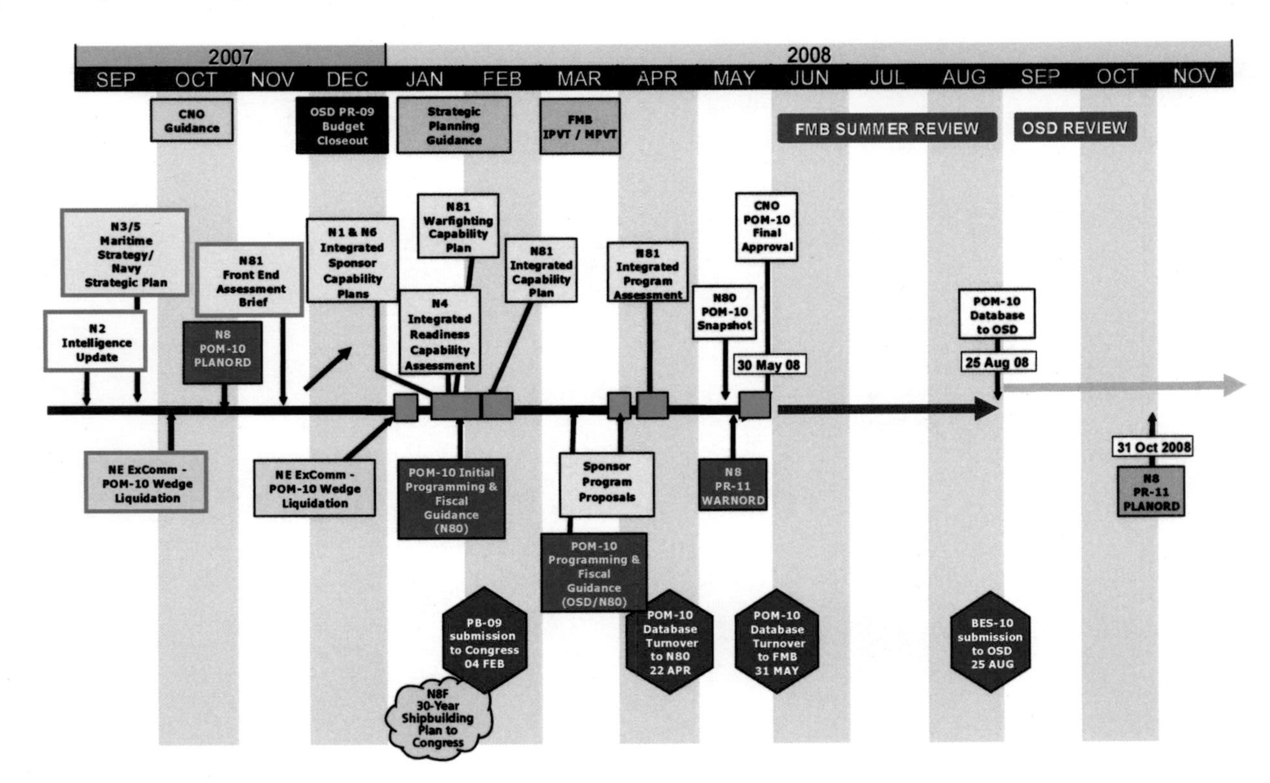

RAND *MG794-3.2*

maker or line of authority for the group of providers, and no clear value proposition (as with the Fleet Readiness Enterprise) for the providers.

Increased communication was identified as the greatest benefit of the Navy Enterprise construct. Under the construct, the drum-beat for regular meetings brought resource sponsors, warfighters, and providers together to discuss money, requirements, resources, and other matters. As a result of increased communication, interviewees indicated that resource allocation decisions and risk evaluation were improved as more and better data and information flowed into the hands of the decision-makers. Many interviewees also perceived that enterprise participation in the wedge liquidation drill was beneficial in helping to assess cost and risk trade-offs. However, two of the participants felt that the time and energy required to participate in the exercise was greater than the benefit of the exercise, as many of the suggestions generated were not used. Interviewees were not able to identify the specific cost or benefit of enterprise participation in the wedge liquidation drill.

Interviewees also reported some adverse effects. For example, some believed that the enterprises did not have enough staff or sufficient programming and budgeting expertise to participate in the PPBE process and were over-extending themselves as a result. Moreover, some interviewees reported that participation of the Navy Enterprise adds another element of complexity to a process that is already extremely complex because of the number of organizations involved, many of which have votes that can delay progress through the POM cycle. At the same time, some interviewees reported that despite enterprise participation, the Navy's PPBE-related processes lacked a corporate perspective, that is, there was no organization that was looking at the Navy as a whole.

Through these interviews, we identified additional areas for Navy Enterprise participation in PPBE, such as an enhanced role in its planning phase. Interviewee suggestions, assessments of individual activities within the PPBE process, and the results of consultations with RAND experts on the PPBE process became the basis for alternative constructs for Navy Enterprise participation in PPBE, which are evaluated in the next chapter.

CHAPTER FOUR

Alternative Constructs

We identified and evaluated three alternative constructs for warfare enterprise and provider involvement in the PPBE process: (1) no involvement, (2) select involvement, and (3) PPBE process ownership. Together, these alternatives offer a full range of levels of participation for initial evaluation. Given the maturity of the Navy Enterprise concept, we do not establish roles for individual organizations within the construct. When we refer to the participation of "providers," we refer to the participation of all nine providers. Likewise, when we refer to "warfare enterprise" participation we refer to the participation of all five warfare enterprises.

No Involvement

No involvement is defined as the elimination of Navy Enterprise participation from newly assigned (POM-10, POM-09) PPBE activities, as shown in Table 4.1. In this option, any recent responsibilities that the warfare enterprises or providers have formally taken on in the PPBE process would be discontinued, and no new formal roles or responsibilities would be given to the warfare enterprises or providers for participation in PPBE. However, they would continue their historical budgeting and execution responsibilities. Those organizations that are BSOs or providers would continue to participate in the PPBE process via their traditional roles. In addition, the historical reliance on individuals throughout the Navy to assist with specific data calls and analysis throughout the PPBE process would continue in a formal manner.

Table 4.1
No Enterprise Participation

Process	Product (Current lead)	Current Involvement	No Involvement
Planning	Intelligence Update (N2)	N	N
	Navy Strategic Guidance (N3/N5)	Y	N
	Initial Wedge Liquidation Proposal (N4)	Y	N
	Front-End Assessment (N81)	Y	N
	POM-10 PLANORD (N80)	Y	N
Programming	Recommended Liquidation Actions (N4)	Y (Provider only)	N
	Integrated Sponsor Capability Plan (N1/N6)	Y	N
	Integrated Capability Plan (N81)	Y	N
	Readiness Assessment (N4)	Y	N
	Warfighting Capability Plan (N81)	Y	N
	Initial Fiscal Guidance (FRAGORD) (N80)	Y	N
	Sponsor Program Proposals (RS)	Y	N
	Integrated Program Assessment (N81)	Y	N
	POM-10 Program Submission (N80)	Y	N
Budgeting	BES (NCB)	Y	Y
Execution	Ready Force (USFF/BSO)	Y	Y

NOTE: Y=participation, N=no participation.

Many resource sponsors indicated that they currently represent the warfare enterprises in the POM process, and this representation would continue. In addition, involvement of the warfare enterprises on a case-by-case basis would be allowed.

Select Involvement

In this alternative, specific activities within the POM-10 PLANORD were selected for enterprise and provider participation.[1] There are many permutations of potential phases, activities, and roles in the PPBE process that the enterprises could participate in—our selections are based on their potential to improve PPBE outcomes or to achieve PPBE objectives. Some of the activities identified in our interviews were

- planning phase (in general)
- future readiness assessments
- wedge liquidation
- front-end assessment
- comment on issues/challenges before the Program Readiness Review
- sponsor program proposals
- review before database is submitted to OSD.

We used interviews with RAND experts, interviews with Navy officials in the Navy Enterprise and PPBE community, and our assessments of individual PPBE activities to develop the selected activities shown in Table 4.2. The warfare enterprises and providers would continue their current participation in wedge liquidation activities. They would also continue to work with the resource sponsors to establish funding priorities, support the warfighting capability plan, and weigh in on the POM submission. In addition to participating in existing activities, the warfare enterprises and providers would participate in a new activity that has the potential to improve PPBE outcomes: an enterprise/provider assessment that would allow them to influence the POM. While the enterprises and providers currently communicate priorities, issues, and needs to the resource sponsors and others throughout the PPBE process, their suggestions for allocating resources may or may not be adopted. By giving them formal input to the process, the resource sponsors and others will be more accountable for their deci-

[1] Note that the majority of interviewees used "enterprise" and "warfare enterprise" interchangeably. In general, these recommendations refer to warfare enterprise participation.

Table 4.2
Current, No, and Select Enterprise Participation

Process	Product (Current lead)	Current Involvement	No Involvement	Select Involvement
Planning	Intelligence Update (N2)	N	N	N
	Navy Strategic Guidance (N3/N5)	Y	N	N
	Initial Wedge Liquidation Proposal (N4)	Y	N	Y
	Front-End Assessment (N81)	Y	N	N
	POM-10 PLANORD (N80)	Y	N	N
Programming	Recommended Liquidation Actions (N4)	Y (Provider only)	N	Y
	Integrated Sponsor Capability Plan (N1/N6)	Y	N	Y
	Integrated Capability Plan (N81)	Y	N	N
	Readiness Assessment (N4)	Y	N	Y
	Warfighting Capability Plan (N81)	Y	N	Y
	Initial Fiscal Guidance (FRAGORD) (N80)	Y	N	N
	Sponsor Program Proposals (RS)	Y	N	Y
	Integrated Program Assessment (N81)	Y	N	N
	POM-10 Program Submission (N80)	Y	N	Y
	FRE Provide Enterprise Assessments	NA	NA	Y
Budgeting	BES (NCB)	Y	Y	Y
Execution	Ready Force (USFF/BSO)	Y	Y	Y

NOTE: Y=participation, N=no participation.

sions to adopt or not adopt enterprise and provider recommendations. The warfare enterprises' and providers' assessments of the POM-10 would be presented during the programming phase of PPBE, around the time that the N81 Integrated Program Assessment is produced. Our interviews suggested that the enterprises and providers largely do not participate in the planning phase, but do provide inputs throughout the programming process. A formal role during the programming phase would help to make those inputs more meaningful.

Process Ownership

The process ownership alternative, shown in Table 4.3, was developed to demonstrate the highest level of potential participation of the enterprises and providers in the PPBE process. In this case, all PPBE management responsibilities (i.e., all N1, N4, N6 and N8, N2, N091, N093, and management functions) would be transferred from the OPNAV staff to the warfare enterprises; providers would retain their current roles. Instead of providing support, the warfare enterprises would become responsible for all capability plans and assessments and for all programming and budgeting. This transition would require a large transfer and realignment of resources (human and otherwise) from OPNAV to the operating fleets and type commanders. Essentially, the POM would be constructed "in the field."

Table 4.3
Current, No, Select, and Ownership Participation

Process	Product (Current lead)	Current Involvement	No Involvement	Select Involvement	Ownership
Planning	Intelligence Update (N2)	N	N	N	N
	Navy Strategic Guidance (N3/N5)	Y	N	N	Y
	Initial Wedge Liquidation Proposal (N4)	Y	N	Y	Y
	Front End Assessment (N81)	Y	N	N	Y
	POM-10 PLANORD (N80)	Y	N	N	Y
Programming	Recommended Liquidation Actions (N4)	Y (Provider only)	N	Y	Y
	Integrated Sponsor Capability Plan (N1/N6)	Y	N	Y	Y
	Integrated Capability Plan (N81)	Y	N	N	Y
	Readiness Assessment (N4)	Y	N	Y	Y
	Warfighting Capability Plan (N81)	Y	N	Y	Y
	Initial Fiscal Guidance (FRAGORD) (N80)	Y	N	N	Y
	Sponsor Program Proposals (RS)	Y	N	Y	Y

Table 4.3—Continued

Process	Product (Current lead)	Current Involvement	No Involvement	Select Involvement	Ownership
	Integrated Program Assessment (N81)	Y	N	N	Y
	POM-10 Program Submission (N80)	Y	N	Y	Y
	FRE Provide Enterprise Assessments	NA	NA	Y	NA
Budgeting	BES (NCB)	Y	Y	Y	Y
Execution	Ready Force (USFF/BSO)	Y	Y	Y	Y

NOTE: Y=participation, N=no participation.

Evaluation of Alternatives

We assessed the potential costs, benefits, and other considerations important for evaluating these alternatives using the following criteria:

- *Workload:* the level of manpower expended by OPNAV staff, the enterprises, and the providers to execute an alternative
- *Other costs:* the level of other resources required to implement tasking, such as cost to relocate staff or additional information systems
- *PPBE benefits:*
  - Cost avoidance or cost savings identified by participating in PPBE process
  - Improvement in resource allocation decisions through identification of cost and risk trade-offs
- *Alignment* of programming and execution
- *Complexity* of the PPBE process
- *Buy-in:* ownership and responsibility for PPBE outcomes shared by stakeholders

- *POM ability:* the ability to coordinate and produce a POM within the Navy and OSD.

Quantitative data and metrics for the above criteria are currently very limited, which makes assigning a value to each metric very challenging. We developed a qualitative evaluation system in which the cost and benefits of an alternative are compared to those of current participation in PPBE using the metrics described below. We used a combination of information solicited through interviews, analytical logic, and available data to assign values to the metrics. In many cases, the value assigned is subject to some debate. Better data and information would be required to arrive at more-detailed or indisputable values.

In this system, the value of a metric is represented in terms of a four-point scale: Relative to current participation, costs or benefits of alternatives will be either better, slightly better, slightly worse, or worse.[2] Workload estimates were not available for the current participation of warfare enterprises in PPBE. Therefore, to assess the extent to which an alternative affected workload, we operated under the assumption that more participation would require more work and less participation would require less work.

Other costs would increase or decrease depending upon the alternative. For example, if the PPBE process was turned over to the enterprises, additional information systems would have to be established, personnel would have to be moved, and other costs would be incurred.

PPBE benefit is the most challenging metric to measure and evaluate; it is subject to debate and requires further evaluation. However, some warfare enterprises and providers provided estimates of the cost savings identified in their wedge liquidation activities. For example, both enterprises and providers identified instances in which their identification of cost and risk trade-offs helped to determine a better allocation of resources. While we cannot measure the extent to which this claim is true, for the purposes of our assessment, we assumed that con-

---

[2] The value "same" was included in the evaluations. However, no option was ever assigned this value. As a result, it was dropped from the discussion to allow for simplification.

tinued participation or increased participation in these and other activities would perpetuate improved resource allocation decisionmaking.

We also evaluated alternatives in terms of the extent to which they enhanced alignment among PPBE phases by bringing together the responsible parties from each phase. Effective alignment is needed to ensure that the products of each phase are consistent and do not conflict. Because the warfare enterprises and providers generally do not engage in long-term planning processes, we focused on the extent to which an alternative improved alignment between the programming and execution phases. Decreasing the participation of the enterprises and providers in programming activities would decrease the alignment of programming and execution because the organization putting together the program and budget may not be informed by the organizations that will be doing the work in question.

Increasing the number of participants and inputs to the process increases the complexity of the PPBE process, while decreasing the number of participants and inputs decreases the complexity of the PPBE process. The current process is highly complex, so more activities and participants are considered worse or slightly worse than the current involvement.

Buy-in refers to stakeholders sharing responsibility and ownership of PPBE outcomes. PPBE outcomes include an executable POM that provides the right capabilities to the Navy and OSD. We identified the extent of buy-in in terms of the opportunities the warfare enterprises and providers had to participate in PPBE. For example, more opportunity represented a higher level of buy-in.

To evaluate the level to which an alternative affected the Navy's ability to develop a POM, we focused on the extent to which the various organizations have worked together historically, as well as collocation issues. The collocation of the various influences (President, Congress, OSD, Joint Staff, and the services) to the POM has enabled an alignment of congressional, OSD, and joint requirements with the Navy POM. We hypothesize that moving the Navy's POM development outside of the Pentagon would hinder the Navy's ability to produce a POM.

Table 4.4 summarizes our assessment of each of these criteria for each of the alternatives.

Table 4.4
Evaluation of Alternatives Relative to Current Involvement

	No Involvement	Select Involvement	Ownership
Workload	Better	Slightly Worse	Worse
Other cost	Better	Slightly Worse	Worse
PPBE benefit	Worse	Slightly Better	Slightly Worse
Alignment	Worse	Slightly Better	Better
Complexity	Better	Slightly Worse	Worse
Buy-in	Worse	Slightly Worse	Better
POM ability	Better	Better	Worse

Pros and Cons of Alternatives

No Involvement

Limiting the warfare enterprises and providers to participating in the PPBE process via the associated resource sponsors and BSOs reduces workload, but it also reduces the current benefits of participation. This already complex process has numerous participants and activities and, over time, it has only increased in both complexity and number of activities. Eliminating Navy Enterprise participation would be a step in the direction of less complexity. Another benefit (with respect to the current participation) is cost avoidance for any personnel required to support a PPBE tasking (e.g., the enterprises and providers would not have to generate inputs, and the OPNAV staff would not have to evaluate them). Additionally, eliminating the warfare enterprises' and providers' participation in PPBE would allow those organizations to focus on execution and on identifying efficiencies in their day-to-day tasks, which may yield more cost savings. The initial focus of the warfare enterprises was to find efficiencies and improve effectiveness in execution, and there are numerous anecdotes testifying to their success in doing so, including lean efforts that have resulted in cost savings. Overtasking the enterprises through participation in PPBE may dilute these potential savings. The ability to put together a POM would remain the same or improve, as there are fewer participants.

Reducing the enterprise participation in PPBE could have some adverse affects as well. If the warfare enterprises are divorced from the planning process, there is the potential for plans to be unexecutable. The sense of ownership and responsibility for PPBE outcomes may also be reduced. In addition, the formal opportunities to positively influence PPBE outcomes (i.e., resource allocation and cost reductions) would be lost. For example, the cost mitigation actions resulting from the Wedge Liquidation exercises would not be available to decisionmakers. Nor would the information and data required for better risk assessments used for cost and capability trade-offs be available as avenues for communication are diminished.

Select Involvement

Select participation—in the activities identified in Table 4.2—would improve on the current participation of the warfare enterprises and providers in terms of PPBE benefit and the alignment of programming and execution. The participation of warfare enterprises and providers in the PPBE activities identified in the table is one way to maximize the benefits of their involvement while minimizing the potential costs. The activities identified were selected based on the potential to improve PPBE outcomes. By participating in the selected activities, enterprises and providers can comment on how resource decisions may affect their organizations. This provides an opportunity for these organizations to identify potentially negative outcomes resulting from current programming and budgeting decisions. They will also have a formal opportunity to help improve the allocation of resources through identifying requirements and assisting with risk mitigation and cost trade-offs. This also gives them the opportunity to identify potential cost savings. The enterprise and provider assessment activity that would occur prior to the program submission would give these organizations an opportunity to influence outcomes. In addition to potentially improving PPBE outcomes, this selection of participation will allow better alignment of planning and execution as organizations involved in execution will also be involved in the planning process. This participation could result in a sense of ownership and responsibility for PPBE outcomes on the part of the enterprises and providers.

Although buy-in and PPBE benefit could improve under this alternative, workload and other costs could increase. The involvement of warfare enterprises and providers in all of the selected activities will require additional work on the part of these organizations, more than currently, as well as the OPNAV staff who evaluate their inputs. It is possible that additional skills and competencies not currently resident in these organizations will need to be developed in order for them to participate on a continuing basis in various PPBE activities. The increased scope of responsibilities and tasking the enterprises and providers will have might offset the cost savings generated through the enterprises' traditional focus on efficiencies and effectiveness in execution. The PPBE process would become more complex as more participants and activities are added, but the ability to assemble a POM would likely remain the same.

Enterprise Ownership

The Fleet Readiness Enterprise's ownership of PPBE could improve resource trade-offs that are currently stove-piped within the various resource sponsors. If the PPBE process is managed by the warfare enterprises, trade-offs between manpower, modernization and readiness could be made by the organizations most affected by these decisions. Currently, these trade-offs are made by N8. In addition, the trade-offs between current and long-term capability requirements could also be determined by the stakeholders who will be most affected by these decisions. The alignment of planning and programming objectives with execution realities would also be improved by such an approach, as would buy-in.

However, under this alternative, costs, workload, and complexity would substantially increase. After the transition, costs and complexity would eventually reach some equilibrium. Considering the amount of time it could potentially take to make such a transition, we evaluate the transitional levels of these metrics. The cultural hurdles and behavioral change required for successful implementation would be significant. The physical implementation costs of such a change would include development of information systems alignment, new reporting structures, new roles for organizations and individuals currently managing the process, and new policy and doctrine. Other consequences, such as personnel churn, would be unavoidable and would have an associated cost. The Fleet Readiness Enterprise and subsequent warfare enterprises would take on significant additional workload and responsibilities to manage the PPBE process. Additional skills and competencies that are not currently resident in these organizations may need to be developed. The complexity of the PPBE process could increase significantly as the transition from one paradigm to the next occurs, and the ability to produce a POM may be diminished as the development would be physically removed from the influences of the Pentagon.

Delegating all PPBE functions to the enterprises has been posited as a return to the former "Baron Structure," in which three-star admirals led the Undersea Navy (OP-02),[3] the Surface Navy (OP-03), and the Air Navy (OP-05), essentially owning the research and development, the procurement, the manpower, and the operating and support dollars of their sponsorship. The Baron Structure had enormous power in the OPNAV staff—each admiral was considered the leader for the Navy in their own right for their area of sponsorship. Each testified before Congress on their various programs; each was the spokesman for their community; and each controlled the day-to-day decisions and recommendations within their sponsorship, such as determining who got the best commands and influencing who became a flag officer.

Enterprise leads correspond to OP-02, OP-03, and OP-05, but they do not own the gamut of resources and, in some cases, are not three-star admirals. Instead, resources are owned by various resource sponsors, and N8F makes decisions regarding the trade-offs among N84, N85, N86, N87, N88, and N89, while N8 makes decisions regarding trade-offs among N1, N2, N4, and N8. Should all programming and budgeting authority be transferred to the enterprises, the current organization would be similar to that of the previous Baron Structure, as the enterprise leads would have larger ownership of much of their resources.

---

[3] The OP-02 shared some responsibility with SEA 08. The Director Navy Nuclear Propulsion, Admiral Rickover, did hold sway over at least the nuclear portion of the OP-02 claimancy.

CHAPTER FIVE

Summary of Findings

As the Navy moves toward an enterprise organization, some fundamental questions have been raised. Specifically, what the purpose of the Navy Enterprise construct is, whether or not the Navy Enterprise is the correct approach to address the evolving goals, what organizations should be an enterprise, and what the roles and responsibilities of enterprises should be, to name a few. Without answers to these questions, the question of the role of Navy Enterprise in the PPBE process may well be ahead of its time. In fact, as the concept matures such a question may need to be revisited.

However, our preliminary investigations based upon the current concept revealed that the formal role of the warfare enterprises and providers in PPBE has not changed much. The enterprises and providers mostly participate in the PPBE process via the various resource sponsors and BSOs, as they have in the past. The single new activities that the warfare enterprises and providers have participated in, such as the wedge liquidation and support to N1, were perceived to be beneficial. The biggest benefit of the Navy Enterprise construct from a PPBE perspective has been the increased communication between resource sponsors, warfighters, and providers, which has helped the Navy to better assess cost and risk trade-offs for resource allocation decisions. However, the additional workload borne by the enterprises and additional complexity brought into the PPBE process could be greater than the benefit. In order to determine this, metrics and data to measure the workload and cost required to execute PPBE tasking, as well as the value of the products produced, will need to be established.

There are many ways for the warfare enterprises and providers to participate in PPBE, each with a set of associated costs and benefits. We constructed three alternatives for enterprise and provider involvement in PPBE that represent the range of potential involvement in PPBE from essentially none to process ownership. While we cannot quantify the costs and benefits, we illuminate the pros and cons of each. These evaluations serve as a starting point for considerations of warfare enterprise and provider involvement in PPBE. Support for our arguments was derived through interviews with numerous senior level officials throughout the Navy as well as from available supporting documentation.

All alternatives can be boiled down to a trade-off between more or less work and more or less benefit. Less involvement will likely result in less benefit, but will also not have the costs associated with current participation. More involvement will have a greater cost, but will likely result in greater benefits. Enterprise ownership of the PPBE process has the potential to be extremely costly and beneficial at the same time. Other considerations such as the alignment of programming and execution, the complexity of the process, opportunities for checks and balances, buy-in, and the ability to produce a POM are other important factors for decisionmakers.

While no single preferred option can be identified in the absence of quantifiable costs and benefits, efforts should be made to foster the benefits of participation observed to date. The current involvement serves as a good pilot for development and evaluation of alternative constructs.

No matter which alternative is selected, all will require senior leadership to drive and direct efforts and activities, define and develop for the warfare enterprises and providers: workload priorities, inputs to the PPBE process, policy and procedure for participation in PPBE, authority and responsibility for activities and deliverables, and goals and objectives. No alternative will succeed if there is confusion over roles and responsibilities or if there are too many initiatives being undertaken at the same time. A successful enterprise transformation will be characterized by clear goals and objectives, a clear yet flexible plan of action, and well-defined alternatives and expected progress.

Bibliography

Assistant Deputy Chief of Naval Operations, "N1 Planning Order for Program Objective Memorandum Fiscal Years 2010 to 2015," memorandum to multiple Navy departments, September 21, 2007.

Buss, RDML Dave, and Mark Honecker, "Navy Enterprise Implementation Briefing," January 10, 2008.

Commander, Submarine Force, "Undersea Enterprise (USE) Overview," undated. As of January 26, 2008:
http://www.sublant.navy.mil/pdf/USE_overview.pdf

Department of the Navy, Vice Chief of Naval Operations and Assistant Secretary of the Navy (Research Development and Acquisition), "Development of a Common Cost Management Framework," memorandum for joint distribution, November 29, 2007.

Department of the Navy, Vice Chief of Naval Operations, "Planning Order for Program Objective Memorandum, Fiscal Years 2008 to 2013," memorandum for distribution, January 18, 2006.

———, "Planning Order for Program Objective Memorandum, Fiscal Years 2009 to 2013," memorandum for distribution, October 19, 2006.

———, "Planning Order for Program Objective Memorandum, Fiscal Years 2010 to 2015," memorandum for distribution, January 10, 2008.

Deputy Secretary of Defense, "Capability Portfolio Management Way Ahead," memorandum to Secretaries of the Military Departments, February 7, 2008.

Navy Enterprise, "Common Cost Management Framework," Web page, undated. As of February 9, 2009:
http://www.navyenterprise.navy.mil/knowledge/tools/ccmf.aspx

———, "Introduction to Navy Enterprise," briefing, undated. As of March 3, 2008:
http://www.navy.mil/navyenterprise/downloads/NE-Intro-Brief.ppt#525

———, "Navy Enterprise," Web site, undated. As of February 9, 2009:
http://www.navyenterprise.navy.mil/#

Navy Expeditionary Combat Command, "About the NECE," Web page, undated. As of February 9, 2009:
http://www.necc.navy.mil/nece/about_nece.htm

Peck, Wendi, *Reaching Across Boundaries to Achieve the Right Results: A Framework for Enterprise Workshop Development*, The Executive Leadership Group, Inc., June 2007.

Sullivan, VADM Paul E., Commander, Naval Sea Systems Command, "Navy Enterprise Background, Evolution and Life in the Navy Enterprise," briefing, New Flag Officer Traning Symposium, August 2007.

U.S. Navy, Chief of Naval Operations, "Navy Strategic Plan in Support of Program Objective Memorandum 2008," May 2006.

The Navy Enterprise has evolved over the past decade to achieve various objectives, from improving efficiencies through lean, six-sigma efforts to producing the workforce of the future. This evaluation of the participation of organizations within the Navy Enterprise in the Planning, Programming, Budgeting and Execution (PPBE) system (1) identifies and describes the current participation of Navy Enterprise organizations in PPBE and (2) identifies and assesses potential alternatives for Navy Enterprise participation. RAND analysts evaluated available documentation and conducted extensive interviews with nearly twenty senior leaders throughout the Navy. The biggest benefit of the Navy Enterprise construct from a PPBE perspective has been the increased communication between resource sponsors, providers, and warfighters, which has helped the Navy to better assess the cost and risk trade-offs of resource-allocation decisions. However, the additional workload borne by the enterprises and additional complexity brought into the PPBE process could be greater than the benefit.

OBJECTIVE ANALYSIS.
EFFECTIVE SOLUTIONS.

$25.00

RAND publications are available at www.rand.org

This product is part of the RAND Corporation monograph series. RAND monographs present major research findings that address the challenges facing the public and private sectors. All RAND monographs undergo rigorous peer review to ensure high standards for research quality and objectivity.

ISBN 978-0-8330-4664-2

MG-794-NAVY